Across the Sky

Michele McCarthy

ISBN: 979-8-9874081-1-7

Quotes from scripture: The Holy Bible: New American Standard Bible: (NASB). 1995. The Lockman Foundation. http://www.lockman.org/nasb/index.php

Dedication

To my Savior and Lord

who reveals Himself everywhere.

Across the Sky

Chapter 1

Study of the Stars

To study the stars, we need to understand the difference between astronomy and astrology. The Merriam-Webster dictionary calls astronomy the scientific study of chemical and physical properties of objects and matter outside the earth's atmosphere. Since God made the stars, astronomy involves the study of God's creation.

In contrast, astrology is based on the supposed influence of these heavenly bodies on humans and events on earth. Astrology entices people to seek stars for guidance, not God, who created them. Astrology twists the truth told by God's stars.

DID YOU KNOW:

- Stars are huge celestial bodies made mostly of hydrogen and helium. They produce light and heat from the stirring nuclear forms inside their cores.[1]
- Our sun is a star. Other stars are light-years away from us and actually brighter than our sun.[2]

1. Stars | Facts, Information, History & Definition (nineplanets.org
2. ibid

Chapter 2

The Star of Bethlehem

The Announcement

The wisemen looked upon the eastern skies

An important sign caught their eyes

Baby Jesus had come to earth

God's brightest star announced the Savior's birth.

Though the subject of this book is constellations, we'll begin with the most important and widely known star of all. The star of Bethlehem.

This unique sign in the sky announced the most pivotal event in history. It alerted the Magi, or wise men, who lived in the part of the world we now call the Middle East, to the birth of Jesus.

Magi are mentioned in the book of Daniel in the Old Testament during the reign of King Nebuchadnezzar. Around 600 years before Christ, Babylonians began taking Jews from Jerusalem into exile in Babylon (today about 60 miles south of Bagdad, Iraq). These scriptures describe magi as astrologers who practiced the ancient custom of getting knowledge from the stars and planets. King Nebuchadnezzar made Daniel chief of the wise men because he, by the direction of God, was able to interpret the King's dream.

Scripture does not mention magi following the star from Babylon but Isaiah 60:6 and 7 declare

> *A multitude of camels will cover you, The young camels of Midian and Ephah; All those from Sheba will come; They will bring gold and frankincense, And will bear good news of the praises of the Lord. All the flocks of Kedar will be gathered together to you, The rams of Nebaioth*

> *will minister to you;* and then verse 9 continues, *To bring your sons from afar, Their silver and their gold with them.*

Gold, frankincense, and myrrh were the gifts brought to Jesus after His birth.

In his book, *Mystery of the Magi: the Quest to Identify the Three Wise Men* Dwight Longenecker points out that after King Nebuchadnezzar, during the reign of King Darius, many of the magi fled to Arabia. Midian, Ephah and Sheba, all mentioned in the Isaiah passage were located in the Arabian Peninsula.[1]

What got the attention of the magi and alerted them to travel west with their gifts? With the advent of computers and astronomy programs allowing a view of night skies in historical times, Rick Larson presents what he believes opened the eyes of the magi to the birth of the Messiah.

Jupiter, the largest planet in our solar system, is known as the King Planet. The first known use of the word "planet" in English was in the year 1300. The word planet comes from the Greek *aster planetes*, meaning wandering star.[2] They were called "wandering stars" because their position seems to move in the heavens'. When Jesus was born, around 3 BC, Jupiter appeared close to the star Regulus, which the Babylonians called Sharu, and the Romans called Regulus Rex, both which mean 'king'. Jupiter converged with Regulus[3] to announce the birth of King Jesus.

Long before the Planet of Kings and the Star of Kings joined to announce the birth of Christ, God had already formed constellations to foretell the life-giving story of His beloved Son.

• • • • ● • ● • • • •

"Where is He who has been born King of the Jews? For we saw His star in the east and have come to worship Him." Matthew 2:2 NASB

"I, Jesus, have sent My angel to testify to you these things for the churches. I am the root and the descendant of David, the bright morning star." Revelation 22:16

DID YOU KNOW?

- Nabaioth was the first born son of Ishmael, Isaac's half-brother. The Arabian Nabateans settled on the east side of the Jordan. It is believed that the Babylonian magi fled to Ephah and Midian in the Arabian desert, presently known as Jordan, during the time of king Darius around 520 BC. These magi were known to be skilled astrologers and priests.[3]
- Regulus is around 160 times brighter and five times larger than our sun.[4]

1, Mystery of the Magi by Dwight Longenecker. Page 69-70

2. Who Were the Magi? | Christ for all Nations (cfan.org.uk)

3. The nine points of Christ's star – The Star of Bethlehem (bethlehemstar.com)

4. Backyard Guide to the Night Sky by Andrew Fazekas, pg. 201

Chapter 3

Across the Sky

God's Plan

Before the Bible was written on scrolls
God named and placed His stars in rows.
He wrote His plan *Across the Sky*
As constellations for children to spy.

How do the stars communicate the story of Jesus?

Imagine a world without computers or cell phones. How would you communicate? How long would it take to relay news to people somewhere else on the globe?

In God's brilliant design, He arranged His story in the stars to be seen by every generation before His message was written. He placed his redemption plan in sight for anyone who looked up.

Do you think God explained to Adam and Eve the gospel message He had posted in the heavens on day four of creation? Did our Creator tell Adam He named each star, just as Adam named each animal? During their walks in the garden perhaps God pointed out each constellation and described each meaning. Then came the fall, when Adam and Eve believed the serpent over God and broke their spiritual connection. When Adam looked at the constellations after he sinned did their meanings come alive? Did Adam and Eve realize God displayed His restoration plan across the sky before they ever sinned?

While we may never know if these conversations happened, God knew many years would pass before writing materials were invented for His Word to be written. Until then, He may have simply taught His people to "read" the stars.

Across the Sky focuses on the twelve best known constellations that make an oval above us. In his book,

Backyard Guide to the Night Sky, Andrew Fazekas points out the Zodiac is unique from all the other constellations because it is positioned on the path of the Earth's orbit around the sun, giving the appearance of a wheel. Over a year's time each constellation comes into view as the earth moves.

This formation, the Zodiac, comes from the Greek word *zodiakos*, meaning circle. The root word comes from the primitive root *zoad,* meaning way, steps, or path. At the core, the zodiac band or circle is *the way* the sun appears to travel in the sky.

> *Jesus said to him, "I am the way, and the truth, and the life; no one comes to the Father except through Me"* (John 14:6).

Malachi 4:2 states, *But for you who fear My name, the sun of righteousness will rise with healing in its wings; and you will go forth and skip about like calves from the stall.* These words give the promise of the coming of Jesus at the close of the Old Testament just prior to the opening of the New Testament where John the Baptist prepares the way for the coming of the Messiah, Jesus.

God's use of words is precise—He never misses a chance to shine truth in unique ways. Jesus, the Son, called the "sun of righteousness" in this passage, could be a fun correlation to the Zodiac circling the sun. God's constellation story revolves around the sun, or better said, *His Son* Jesus, the true Light.

> *Then God said, "Let there be lights in the expanse of the heavens to separate the day from the night, and they shall serve as signs and for seasons, and for days and years; and they shall serve as lights in the expanse of the heavens to give light on the earth"; and it was so* (Genesis 1:14-15).

> *Raise your eyes on high, And see who has created these stars, The One who brings out their multitude by number, He calls them all by name; Because of the greatness of His might and the strength of His power, Not one of them is missing* (Isaiah 40:26).

The heavens are telling of the glory of God; and their expanse is declaring the work of His hands. (Psalm 19:1)

DID YOU KNOW?

- We can more easily observe these marvels (constellations) away from the interference of city lights.
- When you gaze into a clear, dark sky millions of stars appear to twinkle. The sparkle you see is the stars smashing hydrogen atoms together to make helium. They create a nuclear reaction that looks like a burning candle.[1]
- Other countries agree on these images. Places such as China, India, Persia, Egypt, Greece, Rome, and Scandinavia identify the same constellations using names with similar meanings.

Which constellation do you think introduced God's story *Across the Sky*?

1. How can a star burn with no oxygen? (Beginner) - Curious About Astronomy? Ask an Astronomer (cornell.edu)

Chapter 4

Virgo

Christ the Incarnate Son

Many years ago on a pitch-black night
Did Isaiah see stars quite bright?
As Virgo staged the virgin birth
To tell of Messiah coming to earth?

God's story begins with Virgo which illustrates a virgin.

Long before the Gregorian calendar, established in 1582, and followed by most of the world today, the Hebrew calendar marked time. Unlike the Gregorian calendar that marks the new year in January, the Hebrew calendar has several "first of the years". September/October is the time of Rosh Hashanah. Rosh Hashanah means head of the year and is the first of the Jewish holidays and the month when the new calendar year moves to the next calendar year. This holiday is celebrated on the first two days of the month of *Tishrei,* the seventh month of their religious calendar, but the first month of their civil calendar. We begin our journey of the constellations in the first month of the Hebrew civil calendar which corresponds with Virgo.

Have you heard the Bible story about the angel Gabriel visiting the virgin, Mary? Gabriel told Mary she would give birth to a son and name him Jesus. He announced that Jesus would be great and would be called the Son of the Most High and be given the throne of his father David. (Luke

1:30-33). King David, the defeater of Goliath, was promised by God that he and his descendants house and kingdom would endure forever. Jesus is the fulfillment of that promise.

Virgo points to God's announcement of the birth of Jesus. The Hebrew word *Bethulah is* the common name used for virgin in the Old Testament. It represents both the seed of the woman and Christ in human form. An outline of the constellation resembles the form of a woman.

She holds a branch in her right hand. The star marking the branch is Spica, *Tsemech* in Hebrew—the branch. Isaiah depicts Jesus as a branch, *Tsemech,* the branch, refers to the Messiah.

The virgin also holds an ear of wheat, marked by a star, Subilon, in her left hand. Subilon, the star, means ear of wheat. Wheat grains or seeds are mentioned around thirty five times in Scripture. Seeds refer to many different objects in the New Testament—the body of Christ, the Word of God, our words, the church and more. Since this constellation points to the virgin and the birth of Christ, this wheat she holds in her hand likely represents the body of believers.

Scripture states than unless a seed falls to the ground to die, it cannot bear much fruit. An example of that principle is a farmer burying seeds of corn and wheat in the ground and from those few seeds fields and fields of corn and wheat grow. Interestingly, the first book of the Bible, Genesis, calls Jesus a "seed" when God says to Eve that He (God) will put enmity or hatred between the woman's seed (Jesus) and the seed of the serpent. In His mission to save mankind, Jesus died, was buried, and rose again bringing forth the fruit of new life for Himself and the rest of the world. John's gospel declares the seed had to die to bear much fruit.

Therefore the Lord Himself will give you a sign.
Behold a virgin will be with child and bear a son,
and she will call His name Immanuel. Isaiah 7:14

Then a shoot will spring from the stem of Jesse, And a branch from his roots will bear fruit. Isaiah 11:1 (Jesse is King David's father)

Truly, truly, I say to you, unless a grain of wheat falls into the earth and dies, it remains alone; but if it dies, it bears much fruit. John 12:24

I have been crucified with Christ; and it is no longer I who live, but Christ lives in me; and the life which I now live in the flesh I live by faith in the Son of God, who loved me and gave Himself up for me. Galatians 2:20

DID YOU KNOW?

- Virgo is the second largest constellation.[1]
- When we become Christians, we become "seeds", buried like Jesus, so we can bear much fruit. We are a new creation and filled with the fruit of kindness, gentleness, goodness, self-control, love, peace, patience, faithfulness, and joy. Water baptism

is a symbol of burial and rising again.

Where in the sky does God's story take us next?

1. Solarsystemquick.com accessed 10/3/2022

Chapter 5

Libra

Christ the Redeemer

Dangling above in the deep midnight sky
Did young Hannah spy stars up high?
Libra's scales "weighing" the cost due
The high price Jesus paid for our rescue.

Libra showcases scales or weighing.

Have you seen old fashioned scales in the produce department at a grocery store? Customers use them to weigh their fruits and vegetables. They allow an accurate calculation of the item's total weight. Grocery scales allow buyers to judge the weight to assess the correct cost.

The Hebrew name for Libra is *Mozanaim* which means the scales of weighing or scales of justice. In Western culture scales represent the balance of fairness and honesty. In fact, a picture of scales is on the walls of many court houses.

When God made man in His image, He gave them free will. God said man (and woman) could eat from any tree in the garden except the tree of knowledge of good and evil. God wanted man to have a choice to love and obey Him. Instead, when the serpent tempted Eve with the lie that if she ate of the tree she not only wouldn't die, but she would be like God, knowing good and evil, she believed him and ate from the forbidden tree. Adam ate of the fruit as well and sin entered the world. With one bite they gave their authority of the earth over to the serpent and the weight of their choice tipped the scales out of their favor.

God didn't swoop in to reverse the effects of their decision or remove their free will. But He did set in motion a plan for His son to come to earth as a human to take back what the serpent had stolen. Jesus came as both man and God to take back, as a man, what man gave away.

Jesus was and is the remedy for our broken relationship with God. He chose to become a man, live a sinless life, and die for the sins of the world. He provided the way to restore a right relationship with God. God's plan for Adam and Eve to have relationship with Him and authority over the earth was extended to us through His son.

Jesus left His heavenly home to balance the scales of justice on earth for mankind. John 3:16 tells us God so loved the world that He gave His only Son and those who believe have eternal life—a promise fulfilled through Jesus.

Zubin Akrabi, a star in the Libra constellation means The Price of the Conflict. Libra, a picture of a balance or scales, symbolizes Christ as truth and justice, paying the balance due. Man can never balance the scales or buy his own redemption. Christ paid the price and offers His gift of salvation for free.

For the Lord is a God of knowledge, And with Him actions are weighed. 1 Samuel 2:3

Who has measured the waters in the hollow of His hand, and marked off the heavens by the span, and calculated the dust of the earth by the measure, and weighed the mountains in a balance and the hills in a pair of scales? Isaiah 40:12

DID YOU KNOW:

- Libra is visible in both the Southern (autumn months) and Northern hemispheres (spring and summer months).
- Libra is one of the few constellations that illustrates a nonliving or inanimate object.

After the scales of justice, what constellation do you think is next in God's star packed story?

Chapter 6

Scorpio

Christ the Sufferer

While protecting his sheepfold with a sling
Did David spot Scorpio's sting?
Did the shepherd see the stars displayed
The painful wound when Jesus was betrayed?

Scorpio illustrates a scorpion trying to sting the foot of a man.

Have you ever been in a battle? A pillow fight with a brother or sister, a dodgeball game with a group of friends? In the end someone won and someone lost and each person may have bruises or scars as a result. Conflicts imply struggle, fight, and collision.

The constellation Scorpio shows a huge scorpion readying for battle, aiming to sting the heel of a man. This dangerous creature has eight legs and a narrow curving tail.

Scorpio represents the destructive attack of the serpent, and reminds us of God's Genesis promise. He told Adam and Eve the serpent would bruise the Redeemer's (Jesus's) heel. He also said the Redeemer (Jesus) would bruise the serpent's head.

Scorpio illustrates mortal conflict and the many attempts of evil to destroy the chosen seed, Jesus. These stars reveal Christ as the sufferer who gained victory over the serpent.

Powerful people, influenced by the devil, threatened Jesus's very existence both before and after He was born. This epic battle against Jesus began before His birth, unfolding in Old Testament stories. During the time of Moses, when he was sent to rescue the Israelites from slavery, the Egyption Pharaoh tried to kill every male seed (baby) from Abraham. This might have stopped ancestral line to Jesus. Later, during the time of Esther, the plot is revealed that Haman wanted to kill all the Jews in the provinces. This event could also have cut the generational line of Jesus.

In the New Testament, at the general time of Jesus's birth, King Herod heard a new king had been born and fear anyone coming after his personal power. Herod ordered all the babies in Bethlehem, aged two and under, to be killed. Finally, during His earthly ministry, Jesus was scorned, accused of breaking the law, tortured and almost whipped to death before being hung on the cross.

When Jesus did die on the cross, the devil believed he'd inflicted *the* final death sting to the King of Kings. Even though Christ was beaten and bruised as was told in Genesis would happen, His resurrection from death crushed the Scorpion's head in victory.

The name for Scorpion in Hebrew and Arabic is *Akrab* which sounds like "a crab". *Akrab* means wounding, conflict, and war. The brightest star in Scorpio is Antares and means "the wounding". Clearly, Jesus was wounded, but was not defeated. Christ's victory over death brings life to all who choose to trust in Him.

• • • • • • • • • • •

And I will put enmity between you and the woman, And between your seed and her seed; He shall bruise you on the head, And you shall bruise him on the heel." Genesis 3:15

You will tread upon the lion and cobra, The young lion and the serpent you will trample down. Psalm 91:13

A band of evildoers has encompassed me; They pierced my hands and my feet. I can count all my bones. They look, they stare at me; They divide my garments among them, And they cast lots for my clothing. Ps 22: 16-18

DID YOU KNOW?

- In the Southern hemisphere, Scorpio is in the center of the familiar Milky Way.
- *Dschubba* is a multiple star group in the Scorpio constellation. The primary star is a giant—around 15 times more massive than the sun. [1]

Let's find out what follows the scorpion.

1. Scorpius Constellation Facts | Solarsystemquick.com

Chapter 7

Sagittarius

Christ the Conqueror

Did Luke see Sagittarius and know
The arrow pulled across the bow
Aimed to stop the Scorpion's sting
And crush the evil he
wanted to bring?

The constellation Sagittarius forms an archer with a drawn bow.

Did you know people still use bows and arrows? Archery, one of the oldest sports in the world, became an Olympic sport in 1900, but was discontinued, and then reintroduced in the 1970's.[1] Today both hunters and competitive archers use bows and arrows.

48,000 pieces of armor, 10,000 real arrows, and 500 bows were made for the filming of Lord of the Rings.[2] A person who shoots arrows is called a bowman or an archer. The one who makes bows is known as a bowyer. The arrow maker is a fletcher and the bag that holds the arrows is called a quiver.

Sagittarius means *archer* in Hebrew. The archer is represented as a centaur, a creature whose upper half is a man and whose lower half, including the legs, are a horse. His bow is drawn back tight with the arrow ready to fly.

Where does the arrow point? If you guessed the heart of Scorpio, you're correct. Sagittarius aims at the bright red star which marks Scorpio's heart. As described in the last chapter, the scorpion loses his war against Jesus. The devil inflicts deadly wounds on Jesus, but Jesus conquers death and defeats the serpent.

These stars paint this victorious picture.

So, why did God choose a centaur for this archer? One theory is that this half horse, half man represents the two natures of Christ. Jesus came to earth as the God/man to overthrow the enemy.

In the last book of the Bible, Revelation 6:2, Christ rides a white horse, has a bow and a crown and He goes about conquering and to conquer. Perhaps the image of the centaur illustrates Jesus' first coming and victory over death as well as His second coming to conquer as well.

The first four constellations: Virgo, Libra, Scorpio, and Sagittarius, according to Ken Fleming, "portrayed four aspects of prophetic truth concerning the Person and work of the coming Redeemer: the manhood He would assume, the price He would pay, the battle He would fight, and the victory He would win."[3]

And He said to them, "I watched Satan fall from heaven like lightning. Behold, I have given you authority to walk on snakes and scorpions, and authority over all the power of the enemy, and nothing will injure you. Luke 10:18-19

Where O death is your victory? Where, O death, is your sting? I Corinthians 15:55

Gird Your sword on Your thigh, O Mighty One, In Your splendor and Your majesty! And in Your majesty ride on victoriously, For the cause of truth and meekness and righteousness; Let Your right hand teach You awesome things. Your arrows are sharp; The peoples fall under You; Your arrows are in the heart of the King's enemies. Psalm 45:3-5

DID YOU KNOW?

- *Hamartolos* is the Greek word for sinner and means missing the mark—an archer's term.
- Arrows are mentioned 43 times and bows 19 times in Scripture.

- Psalm 45 presents Jesus with arrows pointed at the heart of the King's enemies.

- You can see a decan named Draco near Sagittarius. A decan is a part of the constellation and its message is a part of the sign. Draco represents the great dragon (the evil one) cast out.

The enemy has been defeated, where do we go from here?

1. Archery Facts (softschools.com)

2. 44+ Bow And Arrow Facts - Fact Directory

3. God's voice in the Stars Zodiac Signs and Bible Truth p. 53

Chapter 8

Capricorn

Christ the Sacrifice

Did Paul see clearly in the sky above
That Capricorn stood for God's love?
Dying goat, living fish—joined tight
A symbol Christ's death would bring us His light.

Capricorn illustrates the head of a dying goat with the tail of a living fish.

These stars depict a goat with its head lowered and a leg folded under indicating that he is dying. In contrast the fish tail remains alive and well. At first glance, this unusual combination seems more like a cartoon character than a divinely constructed picture. The Hebrew's called the sign of Capricorn *Gedi,* meaning the kid or cut off. One of the major stars in the grouping is *Al Gedi* which means the sacrifice comes.

In the Old Testament, goats were sacrificed as the sin offering on the Day of Atonement. The Priests killed the goat to atone, or cover for, the sin of the people. This practice had to be repeated year after year until Jesus came to earth, grew in His ministry and died on the cross. Jesus gave His life as the final atonement, the final covering for sin. His death covered *all* sins for all time, ending the need for the death of goats to take care of sin. But this didn't end in death, life came from His death.

Life, seen in the fish half of the image, comes out of or from the death of the sacrifice, the goat.

So why was a fish used to represent life? Could there be a connection between Jesus's first followers and the living fish? The first disciples Jesus chose were fishermen—John, James, Philip, Andrew, and Simon Peter. These men worked their fishing trade on the Sea of Galilee. Jesus invited them to follow Him and told them He would make them fishers of men.

Another interesting connection to fish is that early Christians used the symbol of the Greek letters for fish, *ichthus*, to identify themselves as believers in Christ. While they were under Roman rule, during the harsh early days of the church, Christians were often persecuted, beaten, and or thrown in jail. Because of these dire conditions, Christians used a secret code of sorts—drawing the *ichthus* or fish symbol in the dirt or sand around strangers. If the stranger responded positively to the drawing all was well. If the person didn't understand or respond well, the Christian could remain anonymous and avoid a possible threat.

Capricorn represents life coming from death with a dying goat—the animal of sacrifice, and a living fish—the secret symbol of the early believers.

• • • • ● • ● • • • •

...and not through the blood of goats and calves, but through His own blood, He entered the holy place once for all, having obtained eternal redemption. Hebrews 9:12

And He said to them, "Follow Me, and I will make you fishers of men." Matthew 4:19

DID YOU KNOW?

- In chapter 47 of Ezekiel, God speaks of increasing His levels of living water. This water becomes fresh so that many fish will live. The multitudes of fish referred to indicates the various people who will become Christians. The sacrificial goat dies to bring life to the many fish.

- The Greek spelling of *ichthus* utilizes the first letters of 5 Greek words. The English translation of the five Greek words is: Jesus Christ, Son of God, Savior, or Jesus Christ Gods Son Savior.[1]
- In Latin Capricorn means horned goat.
- Capricorn is one of the least bright constellations.

Where might the living fish lead us?

1. Ichthys, The Christian Fish Symbol Origin and History Facts (biblestudytools.com)

Chapter 9

Aquarius

Christ the Living Water

Did the woman who came to Jacob's well

Know the story the stars did tell?

Aquarius – showed pouring out

Jesus Christ, our Living Water, no doubt.

Aquarius illustrates water pouring out.

Have you ever been dehydrated after running or playing? At one time or another everyone gets thirsty and craves water.

In the book of Exodus, God brought the enslaved Israelites out of Egypt to take them to a place of freedom. Moses, following God's directions, led the newly released Israelites into a desert, where they found themselves parched—completely without water. Even if each person had brought a small amount of water the night of their escape, they'd have run out in a day or two. Imagine the scene of thousands and thousands of people grumbling or fainting from thirst.

God met their need for water by telling Moses to strike a rock with his staff. Fresh water gushed out, enough for all the animals and people.

God provided for those He rescued by miraculously bringing streams of water into the desert as the stricken rock provided life-giving water. Psalm 118 calls Jesus the "stone" or "rock" and the New Testament gospels of Matthew, Mark, and Luke also depict Jesus as the "rock". Did this desert miracle point to the Rock, Jesus, stricken or beaten before His death, who through His death, would pour out living water for each of us? Is the water pouring out in Aquarius a picture of continual living water flowing with eternal life because of our relationship to Jesus?

Aquarius is defined as the Water Carrier or Water Bearer, and represents a man pouring water from a jar. The Hebrew name *Deli* means water jar. Other meanings include blessing out of, Christ the Living, and the place of pouring out.

One day, noted in John Chapter 4, Jesus walked with the disciples through Samaria. He sent His friends to get food, but stayed behind to speak with a woman drawing water from a well. In those days people got water by lowering down a bucket on a rope. Then they emptied that water into a clay jar to take home for drinking, cooking, and cleaning.

When Jesus asked the woman to give Him a drink, she was shocked that a Jewish man would speak to her because she was a Samaritan. In New Testament times Jews avoided contact and conversation with Samaritans because Jews used the entire Old Testament books, whereas Samaritans only taught the first five books of the Old Testament, the Torah. Jews didn't believe Samaritans were pure in the way they followed God. But none of that mattered to Jesus. He came to offer everyone new life.

While talking with her, Jesus shared with the woman many things about her life He couldn't have known if He wasn't God. Jesus explained to her the difference between regular water and spiritual, living, or eternal water. Finally He told the woman that whoever drank the well water would get thirsty again, but He said He could provide refreshing water within—that would spring up to eternal life.

Stunned that Jesus knew her life story, she wanted Him to give her this "living water" so she'd never be thirsty again. Finally, Jesus revealed He was the Messiah, the Christ, the Living Water. At that point she excitedly ran off to tell the whole community she had met Jesus.

How shocked would you be if you were at that well and Jesus told you everything about your life? You can receive the same Living Water Jesus provided to the woman at the well.

• • • • ● • ● • • • •

Behold, I will stand before you there on the rock at Horeb; and you shall strike the rock, and water will come out of it, that the people may drink." And Moses did so in the sight of the elders of Israel.
Exodus 17:4-6

and all drank the same spiritual drink, for they were drinking from a spiritual rock which followed them; and the rock was Christ.
1Corinthians 10:4

Jesus replied to her, "If you knew the gift of God, and who it is who is saying to you, 'Give Me a drink,' you would have asked Him, and He would have given you living water." John 4:10

"but whoever drinks of the water that I will give him shall never thirst; but the water that I will give him will become in him a well of water springing up to eternal life." John 4:14

DID YOU KNOW?

- Aquarius, the tenth largest constellation in the sky[1] is located in an area of the sky often called the Sea because several constellations in that area have to do with water.

How about a move from one fish to two fish?

1.Interesting Facts About The Constellation Aquarius - (astronomytrek.com)

Chapter 10

Pisces

Christ the Liberator

When Peter looked upon the inky sky
Were Pisces, the two fish, up high?
Could he see the world bondage free,
Once Jesus paid sin's price for you and me?

Pisces reveals the two bound fishes.

Books and movies are full of rescues from evil by the good guys. We enjoy stories where the good characters win.

So, who will rescue the two fish bound to the sea monster in this constellation? In the Old Testament, Chapter four of Amos foretells the people of Israel being taken away and scattered with fishhooks. His words came true when the Jewish people were made captives to the Kings of invading nations.

Again in the New Testament times, God's people were scattered around the world as they fled from evil leaders. Jews (and Gentiles) who were in Galilee, Judea, and the surrounding areas while Jesus ministered—and who believed Him—were "freed" from the bondage of their sin and received eternal life. Yet those same followers in the early church, while delivered from spiritual bondage, were persecuted by unbelievers and Roman officials. Some were tortured and some were thrown in jail. Many fled their homes to far countries which helped spread the gospel to surrounding areas.

The "fish" bound by sin and/or slavery became fishers of men. This points back to Capricorn, the living fish coming from the sacrifice of the goat (Jesus). This also connects to the symbol for Christians, still used today, the *ichthus,* which looks like a fish.

Would these scattered people or their descendants ever be brought back to Israel? In Zechariah 8:8 God declares He

> *will bring them back and they will live in the midst of Jerusalem; and they shall be My people, and I will be their God in truth and righteousness.*

God does what He says. Since their dispersion, or scattering around the world, a trickle of Jewish people have returned to their homeland of Israel. In fact there are ministries today who assist Jewish families in returning to their homeland.

God is a God of rescue and redemption who wants everyone free from any bondage, free to come into a relationship with Him. Pisces represents deliverance out of bondage by Christ the liberator.

Pisces is no fishy tale—Pisces is God's freedom trail.

• • • • ● • ● ● • • •

The Lord GOD has sworn by His holiness, "Behold, the days are coming upon you when they will take you away with meat hooks, and the last of you with fish hooks. Amos 4:2

For I will take you from the nations, gather you from all the lands and bring you into your own land. Ezekiel 36:24

And He said to them, "Follow Me and I will make you fishers of men." Matthew 4:19

DID YOU KNOW?

- Pisces, the 14th largest constellation, is 457 times brighter than the sun, but not as hot as the sun.[1]

- Jesus called the men out of their fishing boats to follow Him and told them they'd become fishers of men.

- In 1948 when Israel become a nation again, God made a place for dispersed Jews to return home. People from each of the twelve tribes return each year. Ministries around the world exist with the sole objective of assisting those who want to relocate back home to Israel.

What comes after freedom in the sky story?

1. Pisces Constellation Facts For Kids | What, Where, Size & Composition (nineplanets.org)

Chapter 11

Aries

Christ the Crowned Lamb

Did twinkling stars in the moonless night sky
Show off Aries and catch John's eye?
A gentle lamb crowned in glory
Named Jesus as King of God's great story.

Aries showcases a lamb.

The Scripture or Word declares that Jesus is both the Lion and the Lamb. God's story in the sky continues with the constellation Aries depicting a lamb. The Hebrew name for Aries is *Taleh* and means lamb, while the Arabic name for Aries, *Hamal*, means sheep, gentle, or merciful.

Early in the New Testament, when an angel tells the Priest Zechariah that he and his wife will have a son, they face a big problem. They were older than most grandparents, so they knew they needed a miracle to have a baby in their old age.

During Elizabeth's pregnancy, Zechariah was told his son's name was to be John rather than naming him Zechariah, the family name, as was the custom during those times. Zechariah and Elizabeth believed the angel's words and named their son John. John was the cousin to Jesus and was born several months before Jesus.

John the Baptizer—his title in the Scriptures—was assigned by God to prepare the people to accept the message of Jesus. He urged people to repent of their sins, to turn from their current thinking and way of living and go a different direction—to Jesus—then to be baptized as a sign of their change of life. John announced, "Behold, the Lamb of God who takes away the sin of the world" in John 1:29. When John baptized Jesus, the Son of God, he presented Him as the Lamb of God to the crowds gathered on the shore of the Jordan river. After His baptism, God announced from the heavens that Jesus was His beloved Son in whom He was well pleased.

Some of the individual stars that make up Aries have interesting names. Elnath, the brightest star means the wounded or slain. Sheretan, means bruised and Mesartim means the bound.[1] These names paint the bigger picture of the Lamb as a sacrifice. Jesus came to earth to be the sacrificial lamb for all mankind.

An interesting scientific observation about Aries was noted in Ken Fleming's book, *Voice in the Stars*. In the Old Testament, on the fourteenth day of the Jewish month of Nissan, the sun entered inside the Aries constellation. Moses told the Israelites to take a lamb without blemish on the tenth day of the month and on the fourteenth day they were to slay that lamb in the evening and sprinkle the blood on the doorposts of their houses for protection of the lives of their families. When the blood was seen on their doorposts the plague of death passed over their homes. After this Moses and Aaron led the Israelite's exit from Egyptian slavery.

Remember when Moses led the people to the desert and was told to get water from a rock? The freedom from slavery or bondage and living water foreshadowed, or gave an example, of the freedom and flowing blessings Jesus's sacrifice would bring.

Now fast forward to the New Testament, again, on the fourteenth day of Nissan almost 1500 years later in AD 30, on the possible date of Christ's crucifixion, according to Ken Fleming "the sun once again stood in the head of Aries near the stars Elnath (the wounded) and Sheretan (the bruised.)[2]

On the day Jesus was wounded, bruised and crucified, darkness covered the land for three hours in the middle of the day. Did the bystanders to the crucifixion look up in the blackness of the sky to see the stars of Aires? Did they understand the picture God painted in the sky? God used the same sky sign announcing freedom—first from bondage or slavery in Egypt, then from bondage to sin paid for by Christ's very blood on the cross.

Aries is the last of a group of four constellations that represent the blessings of salvation. According to Ken Fleming, "Capricorn signifies the blessing of life from death. Aquarius pictures pouring out the blessing of salvation's fullness. Pisces signifies the delay of the promised blessing to the Jews. Finally, Aries demonstrates the blessing of Jesus's death on the cross fully realized."[3]

The next day he (John) saw Jesus coming to him and said, "Behold, the lamb of God who takes away the sin of the world." John 1:29

*...and one of the elders said to me, "Stop weeping; behold, the Lion that is from the tribe of Judah, the Root of David, has overcome so as to open the book and its seven seals." And I saw between the throne (with the four living creatures) and the elders a Lamb standing, as if slain...*Revelation 5:5-6

DID YOU KNOW?

- Hamal, the brightest star in Aries is 75 light years away from earth.[3]
- Nissan, on the Jewish calendar, coincides with March/April on the Gregorian calendar-the calendar we follow. It marks the beginning of spring months.

Which constellation follows Jesus as the Lamb of God?

1. God's Voice in the Stars, Ken Fleming page 73
2. Ibid page 73
3. God's Voice in the Stars, Ken Fleming page 71

Chapter 12

Taurus

Christ as the Judge

Gazing upon a night sky glowing full
Did Jude behold Taurus the bull?
His long horns both forward and tall
Reveal Jesus bringing justice for all.

Taurus illustrates a rushing bull.

All of us, at one time or another, have been disciplined for wrong words or actions we have said or done. But have you ever been punished for something you didn't do? Can you imagine being completely without any wrong doing your whole life and taking punishment for every wrong, every sin ever committed, past, present, and future? Jesus took the pain, punishment, and suffering for all of us. Isaiah 53:5 says, *"But He was pierced through for our transgressions, He was crushed for our iniquities; the chastening for our well-being fell upon Him, and by His scourging we are healed."*

When considering the suffering in the life of Jesus, justice sounds frightening. It paints a picture in many of our minds of spankings and revoked privileges. But justice is much more than wrong doers getting consequences and punishment for misdeeds. Justice is right doers being treated fairly—when the judge in court pronounces "not guilty" to an innocent person.

The Rock! His work is perfect, for all His ways are just; a God of faithfulness and without injustice, righteous and upright is He. Deuteronomy 32:4

One definition in Merriam Webster's dictionary describes justice as "the quality of being just, impartial, or fair." Justice is more than punishment for guilt. Justice can be release from a debt or accusation or freedom for the one who did no wrong. It is the mercy of God who gave us His son Jesus to take the punishment for all of our wrong doing. When we receive the free gift of God, we can stand not guilty on judgment day. God's justice, also known as judgment, is always fair and merciful. God doesn't want any to perish, but all to have eternal freedom, eternal life. As God's children we have no fear of the bull.

Taurus, the bull with long sharp horns, faces forward like he is rushing toward something. Some believe Taurus represents an extinct ferocious relative of domestic cattle. In the Old Testament, the power of God for His people is described as being like the rimu or wild ox (Numbers 23:22). Rimu is mentioned in the book of Job asking if that "strong" creature can be bound? The wild bull symbolizes power and rule and Taurus stands as the bold representation of the coming judge and the coming judgment of God.

The book of Jude reminds us that the Lord will come to judge. Understanding God is perfect in fairness and absolutely good comforts us. At the end of his letter, Jude encouraged his readers to rely on Jesus to keep each of them from stumbling. A great reminder to us.

God brings them out of Egypt, He is for them like the horns of the wild ox. Numbers 23:22

Behold, the Lord has come with many thousands of His holy ones, to execute judgment upon all, and to convict all the ungodly of all their ungodly deeds which they have done in an ungodly way, and of all the harsh things which ungodly sinners have spoken against Him. Jude 14-15

DID YOU KNOW?

- In Latin Taurus means bull.
- Taurus is a large and prominent constellation in the Northern Hemisphere.
- A seven-star cluster on the bull's shoulder is Pleiades. Amos 5:8 says *He who made the Pleiades and Orion And changes deep darkness into morning, Who also darkens day into night, Who calls for the waters of the sea And pours them out on the surface of the earth, The LORD is His name.*

What constellation follows the bull?

Chapter 13

Gemini

Christ the King and Savior

When Peter saw Jesus's dreadful plight

Looking at twinkling stars at night,

Did twins Gemini shine a sign

Of Jesus as Prince and Savior divine?

Gemini displays twins.

This constellation, placed in front of the Milky Way is known as the twins. Two bright stars Castor, meaning ruler or judge, and Pollux, meaning who comes to labor or suffer, join in this formation. United, these two represent the dual, or two fold purpose of Christ.

What might Peter have understood about the story in the constellations when he became a believer? When he viewed Gemini, did he feel awed to be part of God's story? When the high priest told Peter to quit teaching about Jesus, Peter and the other apostles explained to the high priest that they must obey God rather than man and that God lifted Jesus to His right hand as a Prince and Savior to grant repentance to Israel, and forgiveness of sins. (Acts 2:28-29, 31)

The Hebrew name of Gemini, *Thaumim*, means the united.[1] Castor and Pollux unite Jesus as suffering Savior and ruling Prince. He suffered beyond our comprehension, yet He is ruler of all the earth. One day He will judge the entire world.

Mebsuta, a star in the leg of Castor, one of the twins in the pattern, means treading underfoot. This points back again to the early promise in Genesis of the Seed of the woman coming to tread underfoot the head of the enemy.[2]

In his book, *God's Voice in the Stars*, Ken Fleming notes an association between the sign Gemini and two close constellations Canis Major and Canis Minor. Canis Major holds the star Sirius, meaning glorious Prince, and Canis Minor holds the star Procyon, meaning the exalted Redeemer and Savior.[2]

He is the one whom God exalted to His right hand as a Prince and a Savior, to grant repentance to Israel, and forgiveness of sins. Acts 5:31

At the end of three months we set sail on an Alexandrian ship which had wintered at the island, and which had the Twin Brothers for its figurehead. Acts 28:11

DID YOU KNOW?

- The star Castor is white, while Pollux glows orange.
- In Greek mythology Gemini represents two heroes in the contest of rescuing the Golden Fleece. The often-repeated story is still available in books.
- The Alexandrian ship that took the Apostle Paul to Malta was named Castor and Pollux and likely had the twins as its figurehead. Ship makers used the figurehead to indicate friendship with sailors and to ward off pirates. A misguided use of ship ornamentation because God is our only protector. However, this gives good evidence that in the first century people knew the constellations.

Where might the twins sail us to next?

1. God's Voice in the Stars, Ken Fleming pg. 89

2. Ibid. pg. 91

3. God's Voice in the Stars, Ken Fleming pg. 91

Chapter 14

Cancer

Christ the Protector

Observing the stars God arranged in space

Did Cancer reveal God's embrace?

Did Timothy see God's safe place

Tending us till we see Him face to face?

Cancer illustrates an inn.

Have you been embraced in a family group hug, held in the middle of a circle of people? Being surrounded by people who love you is a secure feeling.

While some people believe the constellation Cancer paints a picture of a crab with claws closing, other researchers believe the stars form an inn, another name for a hotel. As an inn, Cancer depicts a place of safety and shelter for people when they are travelling.

In Biblical times, inns offered places for overnight stays for people and their animals. An inn was usually an extra room in someone's home. In the familiar account of the birth of Jesus, there was no room in the inn for Mary and Joseph, so they were offered lodging in the stable. Consequently, after Jesus was born He was laid in a manger. A manger was an open box or trough designed to hold feed for horses or cattle.

Cancer, from the root word in Latin meaning to hold or encircle,[1] paints the picture of a group hug or a safe place. The brightest star in this constellation, Tegmine, means holding.[2]

This formation of stars points to Christ as our security and our protector. When Jesus left the earth to return to Heaven to sit at the right hand of the Father, He sent the Holy Spirit to fill us with the presence of God. Jesus wanted us to be held and protected in His absence. The Holy Spirit guides, comforts and protects us. Ephesians 1:13-14 tells us that

> *...having also believed, you were sealed in Him with the Holy Spirit of promise, who is given as a pledge of our inheritance....*

This promise confirms that believers are sealed, or held, until they are home in Heaven.

The Lord will rescue me from every evil deed, and will bring me safely to His heavenly kingdom; to Him be the glory forever and ever. Amen.
2Timothy 4:18

DID YOU KNOW?

- The word "kept" in 1 Peter 1:5 ... who are kept by the power of God through faith is from the Greek word *phroureo*, a military term picturing a sentry standing guard for protection.

- One of Cancer's two clusters of stars looks like a beehive and has 1000 stars.[3]

- A cluster or nebulous cloud can sometimes be seen inside Cancer. The nebulous cloud is named Praesepe, which means manger.[4]

- Ursa Minor, one of the decans near Cancer is known as the Little Bear or Little Dipper. It holds Polaris, also called the North Star.[5]

The final grouping of stars is around the corner, how do you think this story ends?

1. God's Voice in the Stars. Ken Fleming pg. 95
2. Ibid pg. 96
3. Cancer Constellation: Facts About the Crab | Space
4. God's Voice in the Stars. Ken Fleming pg. 95
5. Ibid pg. 98

Chapter 15

Leo

Christ the Victor

The fierce lion resides boldly in space
Leo's tail touching Virgo's face.
His stance confirms Christ's victory
Bringing us full circle in God's story.

Leo depicts a standing lion.

When hard work, either physical or mental results in winning an athletic competition or scoring a perfect grade, the victory feels amazing. Whether a spelling bee, a soccer game, a race or an art contest, success brings a sense of satisfaction. Such accomplishments are worth acknowledgment or celebration.

The most amazing victory of all time is Jesus's victory over death. His victory is a win we can all receive through no effort of our own. None at all. We cannot earn it or buy it. We cannot study for it or train for it. Jesus offers us the gift of a victory He achieved through suffering and death. For free. Our response is to believe and receive the free gift. Leo symbolizes the suffering of Christ is over and the enemy is destroyed. Finally after the enemy bruised Jesus on the heel, as seen through the constellation Scorpio, the enemy is "bruised on the head" as seen in Leo, just what God stated would happen in Genesis 3:15.

There are other interesting facts in the Bible about the Lion. In the Old Testament, Jacob, who the Lord renamed Israel, had twelve sons. Each son headed one of the twelve Tribes of Israel and each tribe was represented by a different symbol. The Tribe of Judah's symbol is the lion.

Not surprisingly, the family of Mary, the mother of Jesus came from the Tribe of Judah. Her husband, Joseph, the earthly step-father of Jesus, also descended from the Tribe of Judah. Jesus connects to the lion in many ways. At His birth, Jesus was woven into the family of the Tribe of Judah, represented by the lion. Upon His return, as detailed in Revelation, the last book of the Bible, Jesus is designated the Lion who has overcome.

Taurus, Gemini, Cancer and Leo are the final four constellations. They represent Jesus as Judge, Ruler in power and majesty, Protector, and Victor.

Leo's tail reaches over to touch Virgo's head, the first sign in the Zodiac we studied, as if to say, mission completed. Everything has come full circle from Genesis to Revelation.

Not only was Leo one of the earliest widely recognized constellations, the brightest star in Leo, Regulus, means heart of the lion.

As we conclude with this twelfth star group, God's story of redemption written across the sky is complete. From Virgo, the representation of the birth of Christ through His life, suffering, death, and resurrection to Leo, His glorious return for us.

All of creation speaks of God's love for us. He named and placed the stars in the sky long before the Bible was written. The twelve constellations that surround the "Sun" twinkle His love *Across the Sky*.

• • • ● ● • ● ● • • •

And one of the elders said to me, "Stop weeping; behold, the Lion that is from the tribe of Judah, the

Root of David, has overcome so as to be able to open the scroll and its seven seals." Revelation 5:5

DID YOU KNOW?

- In Egypt, the Sphinx has the head of a woman and the body of a lion. Some believe this combination represents the Leo constellation since the top of the tail of the lion touches the head of the woman, just as the constellation pictures.
- The big dipper points to Leo.[1]
- The Jews, Mesopotamians, Persians, Turks, Syrians, and Indians each had a name for this constellation that translates as lion.[2]

1. Leo Constellation: Facts About the Lion | Space

2. Ibid

Chapter 16

God's Plan

Redeeming Love

Stars in the sky paint the rescue of man.

Constellations proclaim His plan.

His love for us steadfast and true.

***Across the Sky* God declared, "I love you!"**

God named and placed the stars in the sky. Mythology and astrology seek to distort His message and introduce counterfeit meanings, but truth will prevail. Jesus, born of a virgin, grew to a man, and at thirty began His earthly ministry. He suffered and died on the cross to bring everyone redemption and eternal life. The stars beautifully display how God's creation honors the Creator.

The Zodiac exemplifies Psalm 19: 1-2

> *The heavens are telling of the glory of God; And their expanse is declaring the work of His hands. Day to day pours forth speech, And night to night reveals knowledge. There is no speech, nor are there words; Their voice is not heard.*

Learning the meaning of the constellations in the Zodiac differs from looking to the stars to direct your life. God is King and only He can give us true direction.

Take a few minutes one evening to enjoy the wonder of night sky. Imagine the twinkling stars as God's wink, a reminder of how much He loves you...

After all, He painted His redemption plan *Across the Sky.*

> *I, Jesus, have sent My angel to testify to you of these things for the churches. I am the root and the descendant of David, the bright morning star.*
> Revelation 22:16

Across the Sky

Before the Bible was written on scrolls
God named and placed His stars in rows.
He wrote His plan across the sky,
As constellations for children to spy.

The wise men looked upon the eastern skies
An important sign caught their eyes.
Baby Jesus had come to earth
God's brightest star announced the Savior's birth.

Many years ago on a pitch-black night
Did Isaiah see stars quite bright,
As Virgo staged the virgin birth
To tell of Messiah coming to earth?

Dangling above in the deep midnight sky
Did young Hannah spy stars up high?
Libra's scales "weighing" the cost due.
The high price Jesus paid for our rescue.

While protecting his sheepfold with a sling
Did David spot Scorpio's sting?
Could the shepherd see the stars display
The painful wound when Jesus was betrayed?

Did Luke see Sagittarius and know
The arrow pulled across the bow
Aimed to stop the Scorpion's sting
And crush the evil he wanted to bring?

Did Paul see clearly in the sky above
That Capricorn stood for God's love?
Dying goat, living fish–joined tight
A symbol Christ's death would bring us His light.

Did the woman who came to Jacob's well
Know the story the stars did tell?
Aquarius–showed pouring out
Jesus Christ, our Living Water, no doubt.

When Peter looked upon the inky sky
Were Pisces, the two fish up high?
Could he see the world bondage free,
Once Jesus paid sin's price for you and me?

Did twinkling stars in the moonless night sky
Show off Aires and catch John's eye?
A gentle lamb crowned in glory
Named Jesus as King of God's great story.

Gazing upon a night sky glowing full
Did Jude behold Taurus the bull?
His long horns both forward and tall
Reveal Jesus bringing justice for all.

When Peter saw Jesus's dreadful plight
Gazing at twinkling stars at night
Did twins Gemini shine a sign
Of Jesus as Prince and Savior divine?

Observing the stars God arranged in space
Did Cancer reveal God's embrace?
Did Timothy see God's safe place
Tending us till we see Him face to face?

The fierce lion resides boldly in space.
Leo's tail touching Virgo's face.
Leo's stance confirms Christ's victory
Bringing us full circle in God's story.

Stars in the sky paint the rescue of man.
Constellations proclaim His plan.
His love for us steadfast and true.
Across the sky God declared, "I love you!"

Chapter 17

Constellation Review

1. **Virgo:** The Seed of the Woman: Christ the Incarnate Son
2. **Libra:** The Required Price Paid: Christ the Redeemer
3. **Scorpio:** The Mortal Conflict: Christ the Sufferer
4. **Sagittarius:** The Final Triumph: Christ the Conqueror
5. **Capricorn:** Life out of Death: Christ the Sacrifice

6. **Aquarius:** Blessing out of Victory: Christ the Living Water

7. **Pisces:** Deliverance Out of Bondage: Christ the Liberator

8. **Aries:** Glory out of Humiliation: Christ the Crowned Lamb

9. **Taurus:** His Glorious Coming: Christ as the Judge

10. **Gemini:** His Rule On Earth: Christ the King

11. **Cancer:** His Possessions Held Secure: Christ the Protector

12. **Leo:** His Enemies Destroyed: Christ the Victor[1]

1. God's Voice in the Stars: Zodiac Signs and Biblical Truth, Ken Fleming p. 25

Bibliography

44+Bow and Arrow Facts – Fact Directory Archery Facts. www.softschools.com/facts/sports/archery-facts/787/ accessed 7/2022

How can a star burn with no oxygen? - Curious About Astronomy? Ask an Astronomer. www.curious.astronomer.cornell.edu accessed 8/2022

Cancer Constellation: Facts About the Crab. www.space.com/16870-cancer-constellation.html accessed 9/2022

Ichthys, The Christian Fish Symbol Origin and History Facts www.biblestudytools.com accessed 10/2022

Interesting Facts About The Constellation Aquarius – www.astronomytrek.com/?s-aquarius accessed 9/2022

Leo Constellation: Facts About the Lion www.space.com/16845-leo-constellation.html accessed 1/2023

Pisces Constellation Facts For Kids | What, Where, Size & Composition www.nineplanets.org/kids/pisces-constellation accessed 12/2022

Scorpius Constellation www.solarsystemquick.com accessed 12/2022

Stars/Facts, Information, History & Definition www.nineplanets.org/scorpio-constellation accessed 8/2022

The nine points of Christ's star – The Star of Bethlehem www.bethlehemstar.com/?s=nine+points+of+Christ%27s+star accessed 12/2022

Who Were the Magi?/Christ for all Nations www.cfan.org.uk/search/node?keys=who+were+the+magi accessed 12/2022

Fazekas, Andrew. 2009. National Geographic Backyard Guide to the Night Sky. Washington, D.C.: National Geographic Society

Fleming, Ken. 2012. *God's Voice in the Stars Zodiac Signs and Bible Truth.* Dubuque, IA: Emmaus Worldwide

Longenecker, Dwight. 2021. *Mystery of the Magi: The Quest to Identify the Three Wise Men.* Washington, D.C.: Regnery History

The Holy Bible: New American Standard Bible (NASB). 1995. The Lockman Foundation. http://www.lockman.org/nasb/index.php

Michele is married to her best friend, is a mom and a grandmom. She is a TCU graduate with a degree in education and lives in Fort Worth. She enjoys reading to and interacting with students-encouraging them to live the abundant life Jesus came to give us. She loves Jesus, her family, organizing, scrapbooking, reading, traveling, and all things witty.

Connect with Michele: michelemccarthybooks.com

Also by Michele

Daddy and Me

Aunt Ida Clare

Aunt Ima Mazing

Aunt Iva Story

Across the Sky (children's picture book)

www.ingramcontent.com/pod-product-compliance
Lightning Source LLC
LaVergne TN
LVHW022305200926
848481LV00004B/1225